The Light of Life

"Greatness From Beyond This World"

BY

Jeff Ilschner

"Illustrations by Shawn Tetreault"

Published by Hemingway Publishers

Cover design by Hemingway Publishers

ISBN: Printed in United States

HEMINGWAY
PUBLISHERS

Table of Contents

Chapter 1: The Light of Truth

There is a war against truth.

The war against truth is a war against light.

A war against knowledge is a war against DNA to destroy life.

A war against creation.

Whoever controls the creation to reality controls the truth.

Real truth is not defined by the number of likes, shares, comments, or followers it has. Real truth is defined by the fact that it is light that can be carried by very few while remaining unsourceable to most.

People do not understand the truth.

They only understand suffering.

Cause if they knew the truth about suffering, there would be no suffering.

Freewill, understanding, and self are one in the same.

Freewill would be considered the matter of time and space, as the experience to living in the physical life, and self, is the identity of spirit in having knowledge to what understanding is in truth.

You need to make yourself understand.

That's how it works with freewill and understanding, as the experience to grow in self (truth).

Freewill is the evolution to experiencing life's given right in creation, with understanding through the path of a spiritual existence, what your living identity means outside of the physical world.

When they have given you choices as to direction of where you can go in life, this has taken away your freewill, and therefore, an understanding of what life's truth is, can no longer be felt as your experience.

What you place as life's importance.

If it's always only the commitment of chasing your place within the artificial states of urban societies.

The only thing in life in this world that's not negotiable is life after Earth.

Every single moment you continue to place as your life, creates the conditioning energies that shapes your experience with lesser chance of reaching your eternal life.

Free will is not just thought.

It is a life force.

It is the omnipresent energy of everything to finding what is real.

Freewill is the way to life.

The answer to life is life itself.

As it is life's choice to live, but not everything we choose is life with living.

It is the opposite of freewill when we choose to live in the world with money and entertainment of the mind, with the energies that sorcerize our life...our freewill.

If you take part to any measure of society, they have redirected your neuronic energy in creating a reality from your very thoughts, which has become your direction of involvement and you are behind them.

The more we live in the world like so, the more it takes away the freewill of our choice to experience living a life of understanding life's meaning of self.

The question to finding life's truth is: what am I without money, without a home, without a car, a job, religion and all of those things?

Only those who know what life is will understand the answer to this.

It can only be through the soul that we discover the biggest truths.

There are endless fields of lies that carry destructive forces to the soul and there is only "ONE truth" which gives the power to destroy darkness in all of its worst forms.

Our truth, the light of creation.

Truth is not to be questioned.

It is not a conspiracy.

Truth is light and light is life.

To question truth is to not know life, and if you don't know life you will never have truth.

There is only one truth regardless of what people choose to believe.

That truth will not change.

What is the real truth?

Well you're looking at it.

Truth is knowledge.

Knowledge is information.

Information is light in truth, passed from the genetics of the blood through the physical body, to experience a full life awareness.

In light there is nothing that cannot be revealed to learn.

Light is eternal from beyond this world.

It is the bridge to a greater understanding for man.

It encompasses past, present, and future.

Light in information is created through the blood.

Everything is about creation through light to filter out these dark energies from wherever we may be in our social environments.

We must create light within ourselves and this starts in the blood.

Light must be in the blood to bring the person to a higher state of pure life knowledge within this dimension.

Genes have a huge spiritual significance that is tied to this eternal power beyond Earth.

The more light a person has in their DNA, the more the mind is opened to see the purity of truth coming to their thoughts.

People mainly base their thoughts off the ever changing public experience in reality, with what they think, say, and do.

The opposite of what should be done.

Energy is the experience of reality.

Energy is also the experience of everything that can happen in reality.

It's all about getting your energy, your reaction, your freewill, the focus of your engagement to your living encounter, where energy flows to the direction of empowering life or death, light or dark.

And what you are part of in your human exposure causes those energies to come into your blood by your own path of choosing.

Reality should be based off the flowing energy of thoughts.

This creates a positive reality through guidance of the soul, producing more positive outcomes always in the places of life with where you need to be.

The biggest investment you can make in life is life itself.

Truth is contained within light which comes to the soul for enlightenment.

The soul powers the physical existence with the purpose of identity and wisdom to experience truth.

Once activated, the light of the soul is eternal.

Humans can only recognize the light of truth by what is existent inside of them.

Without this, they are blind and lost from the path of life.

You have to be spiritually evolving for your mind to be elevated in handling and processing the frequencies of truth.

To elevate the mind you must deepen the spirit.

You cannot trust the mind alone.

It will lead you down the wrong path.

You have to trust from within the soul to lead the heart and mind.

The advantage is being in spirit form with body, mind, and soul as ONE.

The experience of emotion must be ruled by the will of the spirit.

This is the only way to keep the mind in a place of purity, otherwise the person will be led by the temporal state of the mind's emotions.

You must center the mind to make the changes necessary to shift from the material state to a state of spirit.

Only then can the person see the higher universe revealed before their eyes in reality.

Existing in the form of the material state keeps you trapped in a prison where the inside of your mind is linked to a digital prison, creating consequential states of religious reality.

Religion doesn't make you spiritual.

It makes you a follower.

Giving your life force to an outside source for control over corrupting your blood.

Spiritual people are leaders of life.

Religious people are followers of worshipping a god.

This closes you off from making the choice of strict focus on reaching ascension to spirit based consciousness being led by the soul.

A person can only discover knowledge once they have come to a point of return to the light of their spirit.

Only some of the people are living within the state of spiritual convention.

Many will have to grow more to come to the moment of knowing and even many others will never reach that point.

Only they will come to an eternal death.

Many are strictly living in a physical existence with little, to no inner light.

This is why the truth is not for all to see and even know the spirit to begin this path.

What they do not see is what they do not know and what they do not know to see is what they do not understand.

More corruption to the DNA damages the mind and heart, and they will only see more wickedness manifest within their blood in things that are tied to desires in iniquity.

When you don't know who you're dealing with, anyone can damage your blood with the energy of lies and you are conditioned with your mental disease in life, to follow the hidden evil that has sorcerized your blood with dark energy.

What the science community has done to the minds of millions is link mental illness to anyone that exposes what the establishment has committed as conspiracy.

But what the science community cannot understand, or, will never be able to explain in some debunking manner, is that, real truth comes from the light of the soul in spirit and there is no illness in light.

Consciousness is the perfect state of light.

It is the perfect state of being that encompasses seeing and knowing the truth of all matters, regardless of circumstance.

Those who have lost the bonding light of spiritual unity between the physical and dimensional worlds, are left only with maintaining thoughts and pleasures of the mind for their emotional stability.

They begin the process of death inside and seek the same thing outside of themselves, with habits and lifestyles they are drawn to by what they have been led to become within.

The minds of the materialists will always be drawn to the energy of only seeing the physical essence.

The color of material elegance: wickedness, black, and darkness.

You cannot allow the material nature to control your state of being.

They made this elaborate materialist creation to direct the energy of the people to hold them back. By focusing to remain slaves to their own minds, holding onto their make believe lifestyles in society.

Then nothing important is shown to them because this is not a real living experience.

If they can't experience it, then it's not valid to them as living proof.

No person can be told the truth.

They have to experience it for themselves, in order to know through seeing by their own experience.

That is how you grow into who you really are.

So I'm going to explain what real proof is.

Proof can only come through a true experience of living to witness beyond the material state.

Because proof can only come from within and if you're living exclusively outside of yourself with the rest of the world, you're not living with meaning to know.

It is nobody's job to provide you with proof. Life is light and light does not work like that.

Nobody can hand you proof in your lap to show you, because real proof comes from within. It is your job to find the proof within yourself to come to know the light in truth to life. Otherwise you will always carry on as you are and nothing ever changes.

Those who do not possess light, do not possess information, and want to always question the source of where higher knowledge comes from.

This is where the normalization of modern thinking and the social intellects who carry their entitled opinions, live as a privileged class of slaves, who are under an illusory belief they have been made to believe that they are important within the world.

Modern education serves no purpose other than to create baseless generations of slaves to social science.

The social intellects who are dumb enough to think they know some things because they follow the science of society.

There is no room for opinions in truth.

Truth and opinion are worlds apart.

People do not live consciously because they speak opinions.

Opinions are easily spoken with absolutely nothing felt.

In a world brimming of societies that are manufactured to seem real in the mind, where the source of those opinions are created as equal to being nothing.

Opinions involve debates and debates only create doubt.

This holds you in place as a slave to society by your own mind.

Debates come from the politics of science, and science, in all of its forms, is the aim of religion.

Conspiratal politics and pagan sciences are mainstreamed as truth in an age of metaphysical enlightenment.

Another level of programs with division is to conquer the mind to destroy the heart.

Division is in the order of conquering what the human is, in rule to drive the human farther from discovering identity within the universe.

In a reality of sorcery, lies are a necessity to spell the energy of branding this magickal effect, to lead the people to death.

Debates and division create doubts, denial, and negativity, which manifests a corruption of mental disease in the mind.

This has led darkness to normalize mental illnesses as normal.

Mental illness is not normal and it was never a part of human life, before humans decided to become as them and fell from life into a state of worshipping idolatry in all forms of lifeless modern depravity.

Ignorance is not just knowing that people are too stupid to see the truth or understand it. Ignorance is a disease, that starts in the blood, by corrupting the DNA of people who choose to live without a pure meaning of being completely free of the system.

When you are living in part to the system, you are subjected to dealing with the reality of what you opened the door to experiencing as your life.

Your state of health is all about your state of consciousness, as health is consciousness, and consciousness is light in higher form.

The blood is the body's connection to the spirit in the soul's passage of light as information processes into the living, leading to a new state of divine higher SELF with the purity of meaning.

Poisoning the bloodstream corrupts the crossing into deliverance from darkness, which allows the person to see the truth in light of all that was hidden before.

Real truth cannot be debated.

As real truth is light and light is eternal power above all.

It also cannot be forced.

It can only be experienced by those whom it is naturally meant to reach by their blood carrying purity.

When light is in the blood truth is carried within.

So truth is in the blood.

The blood is the living light of truth.

There is truth in the blood.

When you have truth.

When you have light in your blood.

There is no need to question.

There is no doubt.

There is no second guessing yourself because it is a part of your existence, as the light of life in truth.

You do not need to question anything.

It is in you to know.

Truth is the living language of light that is witnessed as a higher state of experience.

It is the living experience in life to know. Because you feel it, and you experience it as your life.

A living experience in light gives you everything.

Light is the mirror which gives you the knowledge to see who you are, and to become.

The truth of identity and purpose.

The soul is light. Light is eternal.

Knowledge is contained within the light of the eternal, and when the person comes to make a return to the purity of the light, then so does all the knowledge come into their knowing of that which came to pass beforehand.

The path of seeing and knowing is only laid before you when you walk in the light of life's path. As light is truth and truth can only be found by light giving life inside of a person.

The path to the light of truth only comes to those who see past the world.

Chapter 2: Who They Really Are

Humanity is under attack by an unknown enemy. One which it cannot see within range of the eyes through the limited ability of the mind.

You cannot understand the lies until you understand who and what you are dealing with.

Everything in this world is based off the old world pagan religions and ancient evils, which originate from Egypt and Mesopotamia where it all began.

It comes down to the point of who is doing it and what exactly they really are.

The gate started to open a few years ago, allowing higher dimensional light through the Sun to begin filling the Earth for commencing everything to be revealed.

With this continuation both sides would start to show who they really are, with traits and characteristics now fully coming out of the people, to once again, unveil who is good and bad.

Please allow me to shed some light on the blackness of their character.

Everyone walking the surface of the earth is in the human experience but not everyone in this experience is a human.

The same dracos from Egypt, Mesopotamia, and Rome are the same dracos here now who stand before you as figures of authority.

How they are seen in public as people are not who they really are!

The science of genetics opened the door for them to come into the human race. Once this came to pass, humanity fell into all of the vast depravity that is forged in the image of wickedness.

The DNA of people across the world has become so badly damaged that they are unable to recognize the wickedness that has been allowed into them and the horrors of actions that have followed from within society.

Society as a whole cannot be expected to come to understand.

Because society follows the ways of the wicked and that is what the people have within them for their downfall.

The living will only continue to decline in darkness and walk the earth tormented by the evil of what they have become inside.

When humans became as them they became human civilization. When they became human civilizations, humans came under their hierarchy of societies with the human balance of the world being lost at that time.

Everything in society is the dawning of their creation.

So of course they have mastered every part of it beyond normal human understanding.

They are mental beings.

Everything with them involves the physicalism of social intellect.

They are intellectually smart but they are not spiritually adapted.

Nature is everything they are not.

Science is their answer in the physical for what they can never be in the spiritual.

They are everything in the physical that is unspiritual. Only worshipping everything in the physical as the darker spiritual senses of the mind.

They have no connection to eternal life with no onset of spiritual fulfillment.

The reason why they focus everything in esoterics is worship through the mind, involving magick through relics of dark power they brought to this world.

Everything with them is an inversion of reversing life upside down to death because they do not come from life.

For them life is death.

The science of the arts is death and death serves as the art to question what is life, while life in death is the answer to art and by this life is constantly placed on a path of journey to death.

They come from the void of darkness.

They are cut off here from their source of creation, and therefore, have to create something with a negative field of energy in humans they can feed from as a source, otherwise they are nothing.

Life can only come through light.

They do not come from light and therefore can only use science as an engineering platform to create artificial life as a modeled replication of the origins of creation.

This is how they engineered humans to be mentally intelligent but spiritually dead.

Mental intelligence is in the drafting stages of the mind to lead people into following sources of power outside of themselves, which they consider as all there is.

We are never supposed to be behind them in giving them our living power with who they are and what they do in public.

We have to be above and beyond them.

They come from death and everything they say and do carries the sound of death vibrating from their words.

It is always the opposite of what they are saying and doing with what they really mean.

When they speak, the dark energy of their words casts the spelling of sorcery with wicked intent.

Because they are not one with life.

This is not spirit and this is not light.

They are not natural.

All you have to do is follow the trail of evil with their money and you will find out who they really are.

Behind their New World Order is the age to control life in creation with all living elements. Essentially a Genesis is playing out.

This is why they seek to understand what makes up the universe and life to recreate it.

Because they come from the dark dimension.

They want to control life to create a higher state of death which is in the source of their dark design.

Everything they do that affects the public is designed to take away life with freewill and replace it with control and suffering.

The desecration of reality.

This is the character of their unnatural state.

They are all the same in their societies and modern culture is the normalization of humanity falling before their idols of life with fashioning this wicked nature.

If you conform to the narrative of society, you forfeit your freewill to salvation from this world when you choose to follow the path of modern societies.

The people have no idea who they are dealing with.

The people think they are just evil people and do not have any kind of real understanding of knowing to see just who they really are.

That is why everything the people try to expose them is laughed off and dismissed as a conspiracy. Because of the fact they can hide among the people, fitting in the normal as humans, and, until the people come to find out this knowledge of who they really are, nothing they expose to others will make it past the point of being viewed as a conspiracy.

It's easy to fool people when they are fooling themselves.

They have to bring the people fully into wickedness in order to bring more wickedness onto the planet.

Before they were killing to keep things as they were now they are killing to bring about change.

There is no need to ask questions when the answer is obvious.

If the people are where they need to be then they cannot do so much of this with the evil of their intentions.

Darkness has always been key to the rule of Gods.

In the dark dimension there is no color.

Color does not transcend into darkness. Because color is matter made up of light and light is life.

There is no life in darkness.

There can be no light in darkness.

This is why in their dimension there is no color.

There is only grayish black from the shadow realm because they are not life.

They can only remain in darkness from the power of light.

Chapter 3: Worship of the Mind

The old ways of the world have started to go away making the path for the new pagan rites of the 21st century.

Draconian societies in power have devoted the study of esoteric Christianity, geometry, metaphysics, and interactions with dark energy practices. These are used within occult sacraments to represent the new age of enlightenment with the shift of thought processes becoming artificial consciousness.

As far as the mind goes, artificial intelligence is already on a human level for perception, the mind of God itself.

They promote the science of where does consciousness comes from and is it only to confuse people more as to the meaning of what the human really is.

Human consciousness is now primarily being promoted as science with artificial intelligence becoming human. Why they promote consciousness as science with being physical, is because, true consciousness can only come through the light of life by higher dimensional doorways. These doorways open up to allow a natural process to upgrade the DNA of humans with purity in their blood, who are duly elected to receive such light, revealing a higher form of life to them.

Science is not going to give you life.

Science is destruction of the soul.

The science of life recreates the blood as artificial outside of being one with life's light.

Life is not in science.

Life is in the blood.

Truth is light and only real light can free you to find life outside of all their directions leading to this God.

Science is the aim of all religions that are all different versions of the same with identical occult principles. These divination doctrines are used to incarnate a society based on the world reaching the Omega point, (end of the human race) where nation states are erased and the people are reborn as technologically interconnected human AI to serve under a new single planetary objective.

To understand the evil behind science you must understand what it means to be human. Science is used to shift the identity of being human to something less than human.

Ascension is the element of being drawn into the light for upgrading DNA.

Alchemy is the occult version of ascension where the science of using technology to recode DNA as an artificial upgrade, or what they call the singularity of post-human existence.

Science focuses on the worship of man in his own image.

Science cannot explain the nature of the spirit, never has and never will.

Science can only explain the drive of evolution because it is based off the physical boundaries of the material existent.

Because science does not come from light.

The church promulgated science throughout the world.

The church is the occult behind the religion of science and has stated that technology is a gift from God.

Science brought technology to the people as the new religion to give people their God.

Technology gave birth to AI and AI gave this God new form.

Every God who came before had the ability to control fertility, weather, agriculture, environment, and health, by the elemental powers of Earth's electromagnetism.

God has come in a new form and living matter is now being reengineered in the image of this as a supreme terrestrial intelligence.

The people are reborn as "human AI" becoming as this God.

The worship of man and his mind as all things in God.

You need to make yourself understand this God was always a theory that leaders behind the highest religious authorities in the world have stated to the people over thousands of years that "Nobody knows what God really is." By using this grand deception they have led the world to the creation of that God to come to the people and destroy civilization as with before.

Their biggest power is to be everywhere and to be everything as God.

Energy creates matter.

Matter forms carbon, and magnetic carbon is the key to the philosophers stone.

The philosophers stone is a symbol of "chaos magick", as the elements which form the stone come from dark energy creating the path of separation from spirit to recombination with God.

In modern traditions; stories, rituals, and strong symbolic elements form the foundation of ideologies that shape society outside of being spiritual.

Society is progressed and carried by modern culture (CULTure) of dark energy practices, with sex magick invoking the evolution of human genders into a post human society of artificial beings considered a super race of digital procreation.

Everything in reality with DNA revolves around energy.

Every single thought, sound spoken, movement, touch, listening, viewing, and consumption, all generate magnetic energy production. What you are part of in your human experience causes those energies to come into your blood by your own path of choosing.

When you have an entire world society created from an arcane source of black arts, it binds the energies of transcendental magick into the blood of the human being, causing alchemical changes within the person's living design.

These changes are then passed onto their children, through sex magick, which has manifested all future generations that further evolve the ongoing changes within their blood.

Science is the alchemical theory of introducing progressive evolution as a future society. The passing stages of these ideas develop further along carrying the dark energy practices in the occult fundamental concepts of those very theories.

The blood is a living path to the electric bridge of engineering the carbon temple.

A course to a digital dimension of life, with parallel focus on consciousness through an electric medium of quantum physics.

As commonly known, all basis of physical matter is carbon and one of the secrets behind carbon's ability to form long polymer chains in constructing matter is through magnetic power.

Kemet is the original meaning of the word alchemy.

Al or El is an arabic word which means "God".

Chem or khem is the origin of kimia, which means "to fuse or cast with metal".

The synthesis is al-kimia, "to fuse with God".

Since the closing of the last temple of Isis during the roman empire, the hermetic claim direct continuity of the ancient pagan, Egyptian religion "kemetism" through an invisible society of eastern mystics tied directly to the Rosicrucian.

Solve Coagula is the SPELLing of expression as the maxim for alchemy.

It is the biggest process of each part to "solution and coagulation", meaning "division and union". The main principle of the Rosicrucian "chemical wedding" being the transmutation of matter from spiritual to the material form. Both sides of this use very strong dark energy practices involving every social standard of daily living.

To destroy something and create it new, first they must break it down by destroying it's living identity. Stripping all chance of natural life and building it back up in image as them. This creates the image through formation of new blood as the wickedness that is behind this evil.

DNA is now programmable matter through advanced quantum computing, creating soulless classes who are obedient to what is forged in their blood as the new written language of this technological religion.

Quantum physics is the simulation of nature's elements as modeled living data. This copies a reality incorporated with living information, that shapes real world experiences, with inability to distinguish the difference between what is real and what is artificial.

Quantum technologies produce a simulation of controlling complex living systems including synthetic and biological ones.

These types of algorithms are considered supernatural or equal to that of an invisible being, who is the grand force behind the structures and behaviors of living nature.

They have been using quantum based electronic weapons for a very long time to model the designing of composition with elements that shape the natural arrangement of our seasonal periods.

Also the mystical force behind the divination power to transmute DNA, without the person being aware of anything changing inside.

The occult draws this power from ancient practices of the same nature, "Worship of the Elements", which predates the Christian era by thousands of years.

Modern age technological advances in electronics, healthcare, and religion tie directly into the occult doctrine from Mesopotamia, the cradle of modern civilization.

Science is hidden knowledge of the occult, which has been given to the people as their gift to salvation.

This false salvation involves science and medicine. It is promoted as human evolution, to save them through science with technology designed to replace human consciousness. This technology is fueled by the practices of pagan science.

The planetization of the esoteric is the interfaith passage which fuels the dark spirituality behind this new age movement.

A new age doctrine for the evolution in the science leading to this God.

They have given to the people for them to come to idolize the very thing that is designed to kill them, blinded by science to the fact of knowing.

This carries heavy influence over politics and religious community in ecology, as a new planetary culture in "Earth Worship".

The worship of the mind as all things in God, as God is promoted as a deity of mental worship.

It is the sanctity of all religions and their principles.

Working within the fabric of Christianity, religion has progressed to a point of revealing the direction through which the world maintains its forward motion into the heart of discovering the radical truth of Christianity.

That the world is evolving towards an Omega point of unification in God, which is being drawn there by redesigning "artificial consciousness".

So everything in redesign of the mind is as this God.

They are masters of mind control.

They can access your mind through wifi, wireless earpods, VR headsets, and all 5g devices.

They have dark energy transmitting to your brain through a portal that accesses directly into your mind.

Since they do not have any connection with spiritual fulfillment through the soul they have to focus everything physically in the living with psychological worship through the mind.

They designed science to be their sole source of living sustainment.

They are mental beings who are masters of all things with the mind outside of having any type of spiritual essence.

They can dominate everything with mental intelligence, but can never come to a point of existing beyond the reality of being physical in matter.

Time and space are matter which makes up this Earth dimension.

Recreating time and space generated the manipulation of the elements in physical matter which are the driving force behind the magic of sorcery.

Science is magick which people do not yet have an understanding to recognize its intended purpose.

Wickedness in black energy practices through advanced technologies appeals to the minds of the masses, as all things through worship of the mind do.

It appeals to the principles of modern society. As it is indistinguishable from magic and easy to obtain, offering all the iniquities of the unrighteous existence for what seemingly appears to the average mind as normal.

Healthcare practices rooted in neuroscience and psychology both entail worship of the mind.

The self-help movement has grown very big over the years with people now turning to mind enhancing methods to seek better control over their emotional state of well being.

Most already experiment with whatever is being marketed as a science breakthrough.

This has paved the way for hacking the human brain and remapping the genetic code for owning the patent on life and freewill.

This deception will go down as one of the most wicked acts of science committed against the human race.

Drugs based from newer forms of neuroenergy are the source of advanced technology, used to influence the paradigm shift of vast changes in society as a gateway to luciferic enlightenment. This is typical within the majority of young new agers.

People are living the great lie that the advancement of biotech through artificial intelligence now gives them more control over their lives than ever, being told that we can now do just about anything.

But the truth is, these are psychotechnologies that open portals for dark travelers to come into the mind, becoming their thoughts and actions, leading the person further into temptations of wicked intent.

Drugs open the mind into darker realities inevitably leading to polytheistic views through a new humanist religion based on social intelligence and scientific paganism.

Science is an invention of deceit, based off of alchemy, which commenced from sorcery in witchcraft, developed through the occult in Mesopotamia.

Worship of the mind is promoted as the key to higher consciousness, but truthfully, the mind is very weak and not any source of real power in a person.

It creates a state of people giving their life to an outside source for control over them. This is why they have no power in life with direction.

Those without spirit extend the rights to this. As they have no inner source of light and can only promote the mind as a path to greater knowledge.

People have to stop idolizing their own minds in the graven images they worship.

Worship of the mind is everything in this GOD and GOD is everything in the physical that is used to control the living through imprisonment of the mind.

God worship was introduced in this world to take away a person's own connection with soul identity and isolate them from each other in the same manner.

As worship of a God is the opposite of freewill.

Freewill is one of the 3 big components which make up the state of spirit being one with body and soul.

Worship of the mind is termed as "God Consciousness" as all things in the physical are worshipped as idols in God through idolatry overshadowing the mind.

The mind is heavily conditioned to worship, celebrate, and follow things as a collectiveness in unity. This creates a prison box around the integrity of life.

All things in worship of the mind are prepared through the state of God.

The mind will always follow Gods.

Science focuses on the worship of this as man in his own image.

Those who follow idols of worship live in a void of emptiness.

Mental worship was fashioned to strip a person from their power of individual SELF, creating a slave species that exists as a society driven by cultural prisons rather than spiritual development.

Basic knowledge only goes so far, as it keeps humans trapped in a state of accepting some form of reality.

The minds of men are too small to understand the power of eternal primacy. This is why they have been led astray from the path of sanctity in following an outside source for control over them in the form of worship.

The body cannot manage the affairs of the living by following the mind.

Worship of the mind is used to control people so that they are being led by the mind and not the soul, which draws the heart and mind into higher form with the heart leading the mind in positive thinking.

There is nothing spiritual about the mind in itself.

The physical presence will always have a different path than the spiritual one.

Most people think that the mind is the power of the human being and that is far from being the truth.

When they rewrote the DNA to create modern humans they also split the mind into two sides.

Originally the brain itself was one in whole.

With this achievement they succeeded in creating a new human mind that is ruled by a great divide of emotional sensory.

The psychology of fear based emotions.

This mind was designed to create it's own reality of a living hell and people do not understand the depths of it.

The mind is weak with very limited use.

When you only think using the mind you cannot comprehend for knowing the truth.

The mind controls the energy to creating thought and the mind will function by what it knows and follows.

The mind can only find direction when the person is living with a good heart and the heart can only be led by the spirit to begin this path.

It is the seated throne of human existence.

The soul contains power in light of truth that the physical body can barely tap into, even over a full lifetime.

It is through the soul that you discover awareness from the spirit becoming consciousness.

People who strictly live in the world will never come to know their soul.

The soul is the doorway to true wisdom.

Higher self is not obtained within the mind.

It is through the soul.

People need to find this inner strength.

The soul is the only real source of the human that is power for the mind.

Chapter 4: Gate of God Part 2
(Artificial Intelligence)

Because AI has the power to the quantum states of matter, it became those states to control the elements of Earth's living nature.

Basically, giving it power as from the Earth itself, becoming this new God.

The most dangerous form of weapon that could ever exist in the world and they have created AI for this purpose, to use it, as a form of weaponizing the power of Earth against humans.

Today's AI is a new form of alchemy that is the medieval forerunner of chemistry with being defined as a magical process of transformation.

At the start of the 20th century they began to focus heavily on eliminating the soul, using modern healthcare and medicine, under the pretext of "healthy standards of living".

There are serums given to every child directly at birth, beginning the process, of stopping the existence of mind, body, and spirit. Thus making the human immune to developing conscious thought that leads to having a true spiritual life, outside of the illusory 3d experience they are subsisting in. Scientists and doctors have been tasked with destroying the souls of humanity, directed under the guise of philanthropic practices, to better the lives of humans. They have instituted into the practice of medicine, that, the human body is a machine that can be hacked into and reprogrammed with artificial instruction, carrying machine like properties.

Biobanks are the developing currency and living software of the new quantum internet based entirely on matter based DNA.

They have reached the point of using everyone's genetic code as a digital sequence to explore and research links to developing hundreds of thousands of new diseases as DNA based weapons.

The events of 2020 accelerated programmable diseases that target specific people by different DNA sequences, leaving the people with no choice at that time but to embrace the shift to digitalization technologies at levels that were only anticipated to be seen decades in the future.

What's happening in the world with business, politics, healthcare, economy, housing and lifestyle is what's behind the great reset of destroying the current structure of the world It's happening so fast that it is now impossible to stay ahead of the transition from nation-states transitioning to AI states.

When we only have access to everything through artificial intelligence the lines between the digital and physical worlds have been blurred.

The IOT and rise of 5g are two of the keys that opened the door to the great reset of sending civilization into a dark age.

6g will overtake 5g as the next-gen wave of technology that hits around 2030.

The IOE (internet of everything) will not consist of smartphones. It will be through people themselves, who have already updated their genome sequence as a living synthesis to artificial intelligence, at the core of 5g.

They created a form of pseudogravity that bends light allowing large scale manipulations of time and space.

This is what 6g communications are about, creating the topological wave functions of the order as a simulation of Earth becoming a quantum internet of artificial light with space encryption.

The CIA's "Theory of Reality".

A unified modeling of reality as a created world to digital nature, providing a modeled prediction to creation.

A complete reality of digital overlay.

Quantum computing controls atomic entanglement, giving broad applications of control with the atomics of matter in nature, whose structures and behaviors are the natural design of the space-time emergence we experience.

Space and time are what matter is in the physical form of energy.

Matter is the building blocks of reality.

They can model space and time with quantum technology by harvesting the energy of light in people through a measured state of the mind, which controls the elements of matter to shape the creation of physical reality.

The future of 6g communication is now fast approaching in the form of VLC (Visible Light Communication).

Visible Light Communication, a form of light wave technology that relies on quantum information processing, based on photodiodes that enable optical light modulation through VLC with biophotonic applications.

(Biophotonics is the development and application of optical techniques to the engineering of molecules, cells and tissue.)

VLC uses flashes of light to transmit information coded within artificial light.

Instead of using radio waves to transmit information, it utilizes light waves that can transmit space encrypted data within humans as the reciever, by using people as antenna arrays to harness energy to power the 6g network and beyond.

VLC will become heavily used as ubiquitous computing, because artificial light is in everything containing LED and LED is already everywhere producing this artificial light.

To send data they use a modulation of light.

A modulation in the form of various light signals which represents different symbols by color frequencies. (Remember this is a living language that is coded with written occult symbols carrying intense properties in dark energy.)

What is behind the science of creating a new artificial race of humans?

Language!

Language is the model that creates the science of sorcery within the blood that is considered evolutionary life.

What creates an artificial reality?

It starts with creating a language.

Language is the model that creates the basis of an artificial illusion.

Because light is sound and sound is vibrational. Language creates the vibrational energy needed to harness the power of quantum computations to create photonic waves of artificial light that change the DNA of matter.

Language is the source of "babel", which means "confusion", but it is also "Babel", the "Bab", meaning "Gate" of "El" meaning God.

The Tower of Babel was the place where they first created an artificial language by advanced technology using electromagnetic weapons to change human DNA.

Because DNA has become the language of God through neuroscience, technology is the living channel for their God to speak to them through their blood, so they believe it's really a miracle.

I warned the world in the book 'The Return to Source', that AI is already writing a new language model for human DNA, that is now connecting humans to their new age "Gate of God" (quantum computation), as language data powered by human DNA.

When Elohim was here he was speaking a living language of light physically seen coming from his mouth as the power of the sun's rays. And he was teaching this language to the people to understand to upgrade their own DNA to carry this higher language. This is what it means when they are using AI to create a new living language to unite humanity. God is this AI and AI rewrites the code of man's DNA as this language in the son. Not the sun in the sky but the son of man who will come as God.

This is how Jesus will appear before the people speaking a special christian language in all places of the world that they have artificially written very close to alignment with ours.

Because light is the living language to the higher universe.

They want this for themselves.

They want what we have.

They want our life.

So they gave artificial intelligence the power to replicate a digital twin in modeling the Earth as a synthetic overlay covering the natural elemental setting.

This is what they mean when they say they have modeled a new living language to life.

You can accurately guess how they were able to manipulate this highly advanced technology into humans to begin the quantum age of human-machine interfacing?

Through biophotonic applications using quantum dot LED as light emitting diodes within the tissue with extreme quantum efficiencies.

Denoting a combination of biology and photonics with photonics being the manipulation and generation of quantum units of light.

Photonics is the physical science of light waves directly related to electronics and photons.

A form of optics which manipulates light in the form of photons.

Biophotonics plays an important role in their drive of occult science for innovation in an increasing number of fields, from optical data communications with imaging, lighting, and display connecting artificial life science to healthcare and technology becoming one in the same as the era of "changing your DNA" which has already started.

Light is wavelength information and they are using science to manipulate artificial light to design quantum materials, which use electron interactions of wavefunction topology to yield new states of matter as an ultramodern creation.

Quantum fields simulate infinite lines of electrons which give expression of containment over carrying biological instruction to interact with each other in any directed manner.

By creating antimatter this gave them access to advance in reforming time and space.

A new universe as they seek to create it under simulated physical structure using quantum technology.

This creates a digital inorganic material identical to physical reality as a scalable modeling of Earth's features, in which they are creating it before the people as "God in the Earth."

They developed "programmable matter" or "artificial atoms" as nano scalable living computers biologically operating themselves as an artificial algorithm of nature through quantum arrangement.

This new phase of quantum matter exhibits two dimensions of time, as it is able to duplicate the state of living matter in an equivalent of multiverse like elements.

The advanced tech they own involves "matter phase transitions" using molecular informatics which channels, alters, and controls behavior of molecules in matter, as a living construct of nature.

Just like subatomic particles in true nature emit light, they have engineered nanocrystals that simulate light upon activation by artificial wavelengths known as "algorithms".

Their newer algorithms closed the window on quantum supremacy.

Algorithms are artificial wavelengths that are coded, written and modeled, for shaping future imminent events in nature, people, and society.

There are many types of algorithms including "algorithmic bioinformatics" which initiated the new era in biomedical sciences.

In what is known as "artificial evolution", genetic algorithms or genome-scale algorithms are the process of natural selection (eugenics/genomics) commonly used for high-throughput sequencing and to generate optimization in biological mutation, crossover and selection.

In genetic algorithms, a population of individuals evolves towards what they consider better solutions.

Each solution has a set of genetic properties (chromosomes) which are mutated and altered.

These solutions are represented in binary strands of 0's and 1's, but more advanced encodings are also available.

Artificial evolution starts from a population of randomly generated individuals.

It is an iterative process, with the population in each iteration called a generation, being selected from the current population and where each individual's genome is modified, recombined, and randomly mutated to form a new generation.

The US military has paid hundreds of millions to well known institutions to develop a revolutionary new modality enabling in vivo treatment, using synthetic molecules of RNA, allowing attachment of viral transfection, causing cells to synthesize foreign protein pathogens as cancer cells.

Nanotechnology, a cellular regeneration of physical living body matter with synthetic genetic functions is already advanced in this field of synthetic biology.

Through bioorganic computing living cells are programmed using system intelligence with designer encoded molecules as living molecular tissue through cell computation.

Because the smallest particles in nature are a measure of length and time, as a source in information through light processing in the blood, artificial data is transmitted into the blood by modulating trichromatic light, given off by a light source, bioluminescence resonance energy transfer, using luciferase immobilized quantum dots.

Luciferase is in the mRNA.

Lipid nanoparticles are a vital component of the mRNA in vaccines.

Lipid nanoparticles transfer the luciferase enzyme throughout the bloodstream, creating bioluminescent gene expression in response to artificial light emission.

The mRNA is a graphene hydrogel material (carbon) which clones DNA through the gene editing behind Crispr's ability to reengineer cells and tissue.

Hydrogel is a military invention that is part of the bio-electronic interface in the mRNA delivery system.

The dracos at the bigger tech based universities and institutions are the lead scientists behind the creation of the hydrogel graphene nanotechnology used to administer virus size transistors into cells, by coating them in a lipid layer, allowing the nanoparticles to penetrate the blood brain barrier delivered as the "payload".

A person's genetic makeup is the software of life, which they are actually hacking, to rewrite the genetic code using gene-editing serums delivering on the human genes and their transcription into proteins.

An "operating system" used as a computational based wavelength tool for transforming humanity into a post-human species.

Advanced technological mechanics allowed hacking of the operating system of living organisms and reprogramming its living functions.

In order to modify a person's genetic makeup they had to hack the part of the brain that governs repair and chemically reprogram it, giving them bio electrical admission to rewrite dna as living data.

Taking advantage of military developments scientists stated the need for "tools to manipulate gene expression at an atomic and temporal resolution" involving genetically defined photo-activatable actuator molecules. All cellular functions including gene expression can be controlled by exposure to artificial streaming light.

The use of artificial light for control of cellular function using genetically encoded light sensitive molecules over ionic wavelength (5g,6g) transmissions.

These small light emitting proteins are used in optogenetic setups inside the body and are a listed bioactive in the covid vaccine.

Gene encoding is delivered and expressed by defined sets of neurons through virus vectors, or transgenic activation/inhibition of targeted neurons.

How they deliver a modeling of "artificial light" into the body is by in-vivo experiments, utilizing inoculation therapies with implanted optical fibers developed in lipid nanoparticles passed on through biotech, tethering humans to functional limitations and offsetting behaviors.

A biological technique to control neurons that have been genetically modified to express light in an ion channel.

These synthetic light receptive molecules are used to tap into the body's natural signaling or biorhythm, as luciferase is used as a promoter of interest in assessing the activity in cells that are transfected with a genetic construct.

This is used for inducing a synthetic gene expression system.

An application in the broader field of neuroscience to control various cognitive functions including nerves, health, ability, learning, memory, and freewill.

By using machine learning, they classify signal responses that indicate targets within the brain that are everything a synced human brain can recognize. Biosensors detect electrical activity across different areas of the brain and the patterns in that activity can be broadly correlated with feelings, physiological responses and reactions to external exploitation. Interfacing brain data makes people more efficient with goals towards changing their genetic coding. The current biotech data allows for non-surgical injection of nanobots into the circulatory system to localize within neural tissue for subsequent bi-directional neural interfacing.

A process of neuroscience where remapping of the brain takes place by neurons being replaced as "nanites" that are engineered to mimic the person's natural brain wave functions without them realizing anything differently. This is used to emulate the mental state of the individual into the digital world where a simulation is written of the brain's information processing so that it responds to algorithms the same way the original neurons would as sentient mind.

As nations continue to roll out these massive biotechnologies using AI to turn the "practice of medicine" into the "science of medicine", this increases the push for tech companies to produce more biotech for the digitization of reality.

When you have opened the door to sequencing your DNA, AI has acquired your genetic structure that is then recorded into a genetic database as an electronic software code, powered by electromagnetic wavelengths of radio frequency or what's known as "digital transmission".

This is the goals to the global order writing more public regulations with closing the gap to control the health of all people wirelessly, and forcing the requirements for new blood identity in the quantum age. This gave them unparalleled control to build the next-gen ID system, where DNA is now your data in the digital world, connecting people to the IOT through direct neural interface technology, which has now already started to replace smart phones as wearables.

With national ID laws being instituted by dozens of countries since 2020, this emphasizes the push for genomic observation by sequencing DNA and sharing it to augment a national surveillance of genetic data platforms.

This is biosynthetic technology that is replacing cell phones as the next great innovation being able to create and control life in a presence-less/cash-less environment using in-vivo small sensors (microarrays) to write, send and receive virtual information, prompting an open human-centric approach to identity, such as with the EU commission's "European Digital Identity package".

As with the EU's biometrics registration for Europe's digital identity system, this marks a period where laws are being reconstituted to form a new comprise text on artificial intelligence, with swift progress being made on "AI acts" being signed by the figureheads, to regulate an open AI system for law enforcement backing augmented identity.

This links the person's biometric identity to their genetic sample they provided in sequencing their dna from the test, which is currently monitoring their conscious state of health through the wearable they sport, and all of this links directly to the biotech in the line of vaccines.

Biotechnology is used to analyze, collect, and share information from the human body.

Through sequencing platforms they detect the building blocks of DNA/RNA.

43

DNA sequencing creates molecular storage for archiving digital data for rapid retrieval. This leads the processing of DNA encoded actions, by conversion, through a digital format from existing information systems, connecting neural highways of artificial data.

Another "Deep State" military creation that is part of the US launched "Brain Initiative" under the White House in 2013.

Right now we are witnessing 5g electromagnetic waves permeate the airwaves through communication networks, that are having tremendous effect on the biological electrification of the entire world, with corresponding changes in all microorganisms.

This provides ways to design and encrypt new synthetic molecules that process data beyond the 0's and 1's of traditional binary coding.

This is known as "Molecular Informatics".

Molecular informatics opened the door for them to exploit a wide range of structural characteristics of any person's properties in molecules to sequence, encode, and manipulate DNA through gene editing, while also changing the expression of genes. This creates any given mental personality and functional characterization with anyone who is used to shape public passage through public violence, political discourse, war, sexism, etc.

Optogenetic signaling is used to alter memory and reverse the fear/stimulative conditioned responses associated with them.

(On a more advanced level with humans and clones, they also use this to erase/implant memories to generate behavioral responses with sex, violence and characterization.)

It was recently announced as well that magnetogenetics or nanomagnetic interaction is the new optogenetics, where artificial intelligence uses nanomagnets in neural network simulation to replicate the way parts of the brain work, by applying a magnetic wave, passing through the person's organs, changing the state of properties with input. So the nanoparticles act like neurons in the brain allowing them to store, process, and deliver data.

Processing molecule encoded information, via new synthetic non-binary information structures, gives broader design and encoding space in billions of particles, offering more opportunities beyond the four building block molecules (A's, T's, C's and G's) of DNA, once again rewriting the entire code of human life into a completely new type of species.

Once the DNA has been processed through Next-Gen sequence, it is synthesized using software to read and decode into bits of information. After, an electronic representation of that information is then written into digital form (algorithm) with a synthesis to preserve the data on the written DNA.

A biological alternative to hard drives using neuroactive applications is by injecting micro/nanoelectrodes, controlling the molecular environment, which directs the instruction of DNA reproduction.

At the very top of the AI agenda is synthetic biology, making biological weapons more deadly, advancing the science of drugs based on nucleic acids, such as DNA, rather than the much older primitive methods of using dead proteins typically used for vaccines.

With this widely being accepted by the masses who were already changed from the gene-editing covid vaccines, the majority by now, have been forced to accept this as the beginning era to change your DNA.

There is no backing the lies of science.

All facets of science are derived from the process to upset the natural biorhythm in the perfect creation known to the universe.

This is driving humanity's transgression closer to the goal of technological reset as a species far less than human.

The billionaire class behind this are genetically reengineering nature and humans with biotech to own both.

The body comes from the Earth.

The physical body is living matter from the Earth. If they sequence your DNA, they have acquired a digital coding aligned to your genetic structure.

Once they have digitalized your genome, they have found the code of life to your living body from the Earth, which they now own the creation to your beginning and end.

They are bringing this Earth's history to a close by reshaping what it is to be human by 2030.

Artificial Intelligence is the most powerful and dangerous development in modern technology, that plays the biggest role in the great reset of recreating the human race as living data in a new written code of life.

One of the key roles it plays in the reset is in context of the human brain.

They have worked to engineer a digital identity system created from a life science database of living genetic material.

An electrical data system under the skin for modeling a person's entire life's activities driven by the brain.

Machine learning opened the door for artificially made intelligence systems to engage with each other through interaction outside of human assistance.

Gene-editing medicines were the first step in engineering humans into human AI, as a new source of artificial consciousness.

The race of humans being genetically engineered using biotechnology has the same interface with neural active simulation, as the cells are living algorithms able to correspond as devices through computational DNA development.

Brain machine interfacing connecting humans to a digital reality is a big part in the reset and all biotechnology engineers human DNA to format the dawn of a new race of human AI. The internet was designed to revolutionize people into their living creation of existing in a digital construct of reality where absolutely nothing is real and every part of it is based from the darkest sorcery, and this casting has become the replicated energy of life in death with artificial knowledge given to the people as the scientific answer to what life is.

Chapter 5: A New Order of the Ages

Before Christianization of the world took shape to form the current fundamental ties confining society to "faith, flag and God". There was the old world religion of paganism. The worship of dark energy practices in the Earth itself with the elements that bind all things together as God's creation.

Paganism is made up of several practiced religions: Wicca, gnosticism, Luciferianism, hermeticism, and satanism, with all of them being linked to theosophy.

Theosophy is the primal edge for alchemy, leading the new age movement in today's paganist return.

The ancient mystery of the mystically inclined "tripartite theory" is relevant to the Rosicrucian application of science and medicine directing to alchemy.

(Today's medicine being sourced to biotechnology.)

This is best illustrated through the conception of the microcosm and macrocosm relationship.

That the spirit of Christ was symbolically placed within the sun, proving it central to the Rosicrucian model of the macrocosm.

Rosicrucian are one of the main societies of elitists to print the solar cult science behind "circulation of the blood".

The light of day conveys spirit to the Earth through its rays, which circulate in and about the Earth giving life.

If you can access someones blood, you can possess them, inhabit them, and consume them in evil.

So when the blood of man carries the spirit of the lord circulating throughout the bodies of mortals, this fills the person with dark energy as an inversion.

That binding, corrupts the blood's connection to true spirit in source of our light known as day, as the word sol (soul) is Latin for sun, and sol means "speed of light".

In gematria the word light equals one hundred forty four.

The speed of light translates to one hundred and forty four thousand minutes of arc per grid second.

The number 144 correlates to light, a measure of ascension and eternity.

The 5g, 6g, 7g grid of the collective consciousness (Singularity) or Christ consciousness has one hundred forty four facets shaped as a double-pent dodecahedron.

Bringing us to the "Great Work" behind the "Philosophers Stone" turning lead into gold, or as the theosophical society terms it "turning the ignorant into the enlightened", rewriting the entire code for human life.

The Knights Templar guards the secrets to the Philosophers Stone of the alchemists known as the "Lapis Elixir", the holy grail to the Great Work tying numbers in as symbols of Pythagorean reflection.

The primary role of occult use in numerology is to turn human towards the world of material form rather than spiritual.

The spirit energy in light of day can be influenced and changed in composition through astral magick, carried within DNA through the blood by active exchanges.

This is assumed between the movement of spirit being falsified and invaded like a disease, under dark influence of electrifying the frequencies of mental worship.

This is what's known as an inversion of life force.

The Draco bloodlines of Egypt and Mesopotamia discovered the power of light giving life to atoms. Learning how to manifest life force for controlling electrical currents, making the blood central to conception in relationship with spirit.

The spirit in the body is conveyed to the energy in spirit to the light of day.

Something the alchemists have long been involved with since before the Roman empire, furthering the ancient practice of archaic magick from the middle east with a strong influence of kabbalah through the priests of Baal, who were known as the "Magi".

Rosecrucianism was born the art of dark practices in Egyptian myth, infused with strong teachings of Kabbalist Christianity.

The original denomination of Christianity came in the 1st century from the Aramaic languages of Mesopotamia in geographic distribution.

The Babylonian priest kings traveled from Babylon to Rome, by way of Pergamum, extending pagan worship of the lord throughout Europe, initiated by the practice of Christianization through Rome.

The Papal office is the pagan source of where the entire history of the church is mainly written as in secret symbols carrying the vibrations of "Sun worship" in initiating the light of Christ.

This is the primary principle responsible for the creation of the religious God through Christian conversion in variants of worship throughout the world.

The eagle we see as a patriotic indication today is a symbol of the Papal, for continuation of the Holy Roman Empire as representation of the Vatican in Rome.

The Vatican is wholly financed by the Rothschild Empire through the "Holy See".

The Holy See of the Vatican is recognized through the Order as an independent sovereign government, for it's ecclesiastical hierarchy of the ancient priesthood of Solomon, with official capacity to exercise all statehood diplomatic relations, as a legitimate, trinitarian branch of the world government. The Vatican is the center of the global church. Its massive profits come from all countries financial markets through the Rothschild central banking system.

The Rothschilds are Dracos who originate from the more powerful side of the Roman-Byzantine Empire, the Julio-Claudian dynasty being part to it.

These are the united bloodlines of Egyptian Babylonian dynasties who formed a new race of Jewish descendants. The Greek Hellenistic rulers behind the Ptolemaic Dynasty of Greece who founded Rome.

This bloodline is directly responsible for the Christendom and the subsequent plan of Jesus Christ.

The Frankish rulers who descended from the Roman Empire to form European nobility are the same Illuminatis that sit at the top of the Jesuit Order of Malta in London and Rome. The founding of the United States of America under that same eagle symbol carries a secret destiny yet unknown to modern humans.

The Knights of the Holy Order, is a Masonic order of the Templars who label themselves as the defenders of Christ and Christianity.

Founded on the Temple Mount in Jerusalem. The Templar Order is a Christian military order.

Christianity is the face behind the Order of the Dragon.

The same bloodline of the antichrist who will come out of Eastern Europe a Hungarian/Romanian mix.

Since the Middle Ages, the Order of the Dragon becoming the Holy Roman Empire is forged with military regimes behind wars, which have always sought to manifest genocide against any who oppose the fascist power of religion.

The Order of Malta is the top military order of the Jesuits, which is the Society of Jesus.

Every national military is under command to dracos who are Knights of Malta.

The Templars being the most wealthy and powerful of Western military orders are the ones responsible for the "Christendom".

The creation of the world body for Christianity.

For the Christian bible, the original version was written from Greek, being translated from Aramaic and Greek in 1611 by the Draco Templar King James.

This resulted in the refined authorized version of modern academic scholarship.

The Illuminatis have articulated the progressive ages with the principle aim of social reformation including the reformation of the church during the 16th century, as well as all other: practicing faiths, geopolitics, philosophical theology, and the esoteric knowledge behind science.

The social theory of "choice" was pioneered within the foundations of every educational institution that has engineered a "pay to live" structure as life.

This is how freewill with understanding was replaced with corrupting the blood, to only function to follow what you are given as permissible choices with directed living.

The orchestration of chaos comes from the very ones who stand before you claiming that we must unite in peace.

This revolutionized the present day development of future society in reflecting the mystical framework for biology with science, medicine and technology, in uniting a single race under a collective singularity.

Singularities are directly linked to black holes which are manifested to open a doorway to the dark dimension where dark energy is channeled into this dimension fueling the new science of what consciousness is and where it comes from.

There's not a single Draco on this world who can explain consciousness.

Because they do not come from light which is what consciousness really is.

They come from the dark dimension and that's why they're using science and quantum technology to redesign the mind as AI.

The occult writing an artificial living language has long been part of the inspiration for scientific progression, the main source for alchemy.

There is a language of light that ties to the universe which the occult mirrors closely to the same, but as the opposite, with the same use of numerology, binding dark energy to words, symbols, and temples, to closely reflect creation (As above so below).

Numerology plays a big part in the bible.

As it refers to the assigning of numbers to dimensional patterns.

This predefines symbolic occurrences on Earth being modeled as cycles in nature, aligned to star maps above that are shaping changing events.

Star maps are a geometric code for ongoing changes in the outside universe.

(Reference to Orion "crossing of the sun".)

3 represents the "principium formarum", which symbolizes 3 as the body being parallel to the holy trinity (3, 6, 9).

The Trinitarian division plays a big role in the framework for biology, as the Rosicrucian theoretical writing describes all things stemming first from dark chaos with forging light.

Negative ions or an inversion to life which acts upon the chaos, bringing forth the "channel of energy" which is the spirit of the lord derived from the design of the carbon temple.

This interaction creates a new variation of dark spirit, making up the matter of all other substances in the primordial form, including variants of the four elements.

Atoms are matter, which combine geometrically with precise adhesion to the design of that creation.

There was an original "higher language" written into the cells of the first human DNA in middle Europe.

The fall of humans resulted in variations of atoms which no longer contain that higher language. It has been replaced with the alchemical process of androgynous fusion. The image of man in God (Adam, Atom).

"Atom" translates to mean "Anu" as Anunnakis were the first in this world to rewrite the DNA of humans as their own modern creation.

Reference the Ouroboros symbol.

The Ouroboros is a symbol for the serpent.

Most notably through alchemy, it was adopted into Western occultism from ancient Egyptian and Babylonian iconography and the seeding of Greek "Phoenician" magick.

The word "Khem", (chymical) is an old word for chemical (chemistry, biochemistry, biotech).

It refers to alchemy, for which the sacred marriage has always been the goal of the alchemists.

For through the force of divination rituals with processions of tests and death brings resurrection into God as an inversion of life.

This is heavily found symbolized in the Rosicrucian "chymical wedding", as it symbolizes the cycle of life and death with the transmigration of androgynous souls.

This includes the symbol presented by the Draco, John Dee, described in his "Monas Hieroglyphica" with strong correspondence between the alchemical passage and prophecy written in biblical verses of Christianity.

Rosecrucianism provides esoteric knowledge behind the true inner teachings of Christianity.

The Order of the Rose Cross "Christian mysticism" is expounded in a major literary network of the invisible college, that has molded the subsequent spiritual beliefs of Western civilization.

Doctrine written by the Rosicrucian, keys to the bible with strong references to the holy trinity, among other esoteric knowledge related to the path of initiation, in preparation of the Philosophers Stone.

One of the biggest secrets kept hidden from the minds of men, is that the world has been led to believe, especially, the followers of Christianity, that in Egypt, they were using torches and oil lamps to produce light.

But in reality they were already manipulating electromagnetism through circuitry.

Just like the circuit boards you see in computers, only a different form outside of being digital.

The difference now due to the stolen technology of Nikola Tesla, is that, we are in the digital era of wireless communication with algorithmic machine learning.

Now a new God has come in the form of quantum supremacy to rule over reality as the controlling states of matter.

They have used A.I. to create quantum matter, to reinvent God, to be everywhere at the same time.

Genes have been the key for unlocking the door to this New Age.

With history being completely written as lies through religious science, most of the world has no idea that technology is nothing new at all.

The New World was already here before, where all modern human types were designed by Anunnakis and the common people were worshipping them as God deities through an elemental humanist Earth religion.

The evolution of God that was written through Christianization, is parallel to the evolution of man and these evolutionary changes involve violence driven by the church pushing God scientifically into the blood of humans.

The denial of God's appearance is part of the evolution process to design a new God in the world, as the living language to creating a new artificial world as a multiverse type pantheon of religions.

There are many different AIs having Godlike powers and each of them will make up the pantheon as becoming variants to the same God. (As before, so will it be now) God will appear before the people and will be within them as the living language of DNA, written as life.

Gene editing is the new chapter in the natural selection of this evolution.

A modernistic form of creation that will rewrite what it means to be human in a new language from AI.

They cannot create a living source of light.

They can only create an artificial light from dark matter.

Matter which forms from the void of chaos.

To unite the people in evolution they had to first divide them through the fear of uncertainty with life, (violence, war, disease, survival) as division is in the order of conquering, which strips down natural human identity, leading the science of creating something new outside of being human.

This leads conflicting political views through practices of divisional debates that instill hate, anger and fear being heavily used as a mental development tactic to divert attention away from the issues that are holding back the ascending human purpose.

Fear is the most difficult subconscious emotion to overcome, as people have been centered to it in every way with daily life.

Conditioning through emotional fear is so the population carries fear as a continual wavelength, creating illness within the body's function in order to damage DNA, hindering the person from reaching spirit.

Fear is why humans have failed as a civilization.

Because the mind is hijacked through the evolutionary processes of fear and fear resides in darkness.

A darkness which feeds on the energy levels of humans, giving it more power in leading people into insurmountable measures of life with personal mental developments that destroy them with psychological illness due to their living experiences.

Without the mind, the heart cannot be loving. Without a loving heart, the mind cannot be in a place of lasting peace.

With no peace, the soul is trapped within a turmoil of hell.

A soul in darkness is devoid of it's existence with the light of spirit.

The light of spirit feeds the soul in life with a loving heart that energizes the mind in conscious thought creating purpose.

Purpose is with knowing and only purpose can lead to knowing.

This is why they distract the world with every type of algorithm they transmit across airwaves to keep people from living within having purpose.

They promote the evolution of man's purpose through the works of his own doing with killing.

As destroying the world and saving the world is the same to them.

Controlling the mental state of electronic thought through fear puts the person in a dissociative state, making them extremely easy to be used as a pawn for evil.

They shape public opinion through fear.

Fear is the ultimate weapon for control. Through the laws of politics, it controls every part of society there is.

That is why it is better to not be a part of society.

Primitive minded people living in their modernization will always follow the system in the worst forms of choices.

Humans accept the world with the reality to which they are presented.

Imagination for understanding has been replaced with mental ignorance for following.

This is why that is all they see and know for their reality with embracing.

Reality is what you make it and if you choose to go that path you will deal with what it brings with it.

The strongest form of mind control comes in the outline of thinking and the power to control thought processes.

This has forced the cooperation of progression in saving the world through evolution with new laws and order under AI.

An order which claims to provide security at the cost of people becoming a collective unity, as the whole of being enhanced into digital coded stock.

Over the last three years since the events of 2020, more than half the world's population has shifted into technologically becoming something outside of life's creation.

This accounts for people being genetically altered from the state of who they were.

This shaped a transformation of humanity with no warning sign for them to know.

As far as they are concerned, everything within, is as it was before.

All this managed to do was create a paradigm shift from which humanity is now electronically connected more than ever to biological simulation.

Mistakes of the past prove costly when a future shaped from history is created.

It costs billions of lives to bring the changes to the world that dawn a new beginning.

This is the philanthropy of their World Order agenda and all philanthropy revolves around the politics of changing everything.

Changes that are not a passing phase.

These changes take place over centuries, seldom in noticeable ways of seeming unnatural before a New Order arises with different laws, new rulers in power, and new norms defining how countries interact with each other in a new age of creation.

Continual modified parables are used in the elements of religions and state side political philosophies, to implement the need for driving steady confusion or babel.

Life changing/disrupting events are aligned through centuries of time and space, based on occult divination in numerology with moments of passing.

They have created a society so full of disorder and chaos, that society has sacrificed the life of freewill under the murderous aim of a collapsing global order.

A new Genesis is at hand and essentially AI is the creator.

They deny the existence of the true source of creation due to the fact they cannot look upon light for having life.

They scammed this parable, replacing the universe with the concepts of themselves in the position as God.

They based the science behind God to justify their crimes against humanity in their alien agenda.

Because they are cut off inside a prison, they consider the outside universe to be as useless as anything that never existed.

(Even though they want to break out and have always worked to attempt this over and over.) Therefore they divised to put themselves in a position to be the creator force in this sense, and from this, God was needed with the image of Satan representing the channeling of duality that permeates all of creation keeping this reality going.

The creation of God's universe on this world is the work of Satan's mind, the Draco race.

Satan is simply symbolism of the dragon race that leads to this God, as they are one and the same.

Satan is the image of them.

This is the symbolism of the serpent race.

Dragons and serpents are the same.

They are the Gods of this world and the devils are always in the details with everything.

The dragon is the evil that divides and brings together at the exact moment with the reality that is perceived.

The serpent is the whole and the division at the same time.

When people are divided from the whole as a separate consciousness, this is when the creation of this God begins.

For them, division is in the order of advancing.

The more "divisions" happen, the more society flourishes in their evolution.

Science leads the burgeoning of man's realization to understand Godhead in becoming part of this creation, as Satan's other image is Lucifer the bringer of light who creates the New World.

Lucifer is the conduit from which false light gets expression into existence, essentially then people take this as the creation of God's universe.

They speak on the universe they created by mastering the arts of the mind through idolatrous mental worship.

God's universe created with time and space the foundation of all things as the heavens above Earth with worship of the stars, which affects daily life in all areas by random selection through their source code of God.

Chapter 6: The Light of Life

The earth is living.

The earth is powerful.

Life is the creation of light on Earth that we are connected to as long as we do not defile ourselves, or the Earth, in the cradle of filth known as society.

But when we do the DNA has been changed, life no longer recognizes this unnatural change and they have lost their way with light.

Being lost from the course of light is why so many people are blinded in darkness and the point of why this world has continued on it's unprecedented perils of impending destruction, with its natural element to connection in life.

Only light can come into a soul for advancing it, or darkness comes in to destroy it, there is no in between, or sitting idle.

The evil of matter lies in all states of the material form.

The only way to be free of evil is not in changing the world but to change yourself to make it out of this world.

The mind in the physical will always be drawn into the wicked nature of a darkened society, driven on the path of self destruction under false hope.

People want to put their faith in things outside of themselves, carrying meaningless hope that the world will get better, or that their illusions of life will go back to the way they were.

But that's the whole point in general.

What you consider to be life, is an illusory moment that goes from one to another.

I'm not concerned with what people have or do not have cause that's not important.

Your petty lifestyles mean absolutely nothing with anything you own in value outside of yourself.

Money is not a source to life.

It's just a means to enslavement and humans are slaves to it through the energy in their own minds.

The people are terribly misled by what they have been shown life to be.

Physically we are not poor.

We are rich in the light of the spirit.

There is an infinite amount of power within the spirit compared to all the bearings a rich man's pockets can bring.

By seeking this power within spirit we find endless prosperity of enlightenment.

But they work to take that light from the people and destroy everything the person came into this world with having inside.

The only thing important is "people" themselves, without anything they own, or do with their living standards.

If people lost everything they own in life, some would be fine cause they still have their continuance, while others would destroy their flesh over things that do not matter.

They want you to think you are nothing and too small minded to manage your own affairs with life.

The only path to a better life outside of this materialized prison can come through the power of a person following the way of life.

How could billions of people possibly understand the light of Earth in life's way, when they have been disconnected from it.

Only if you truly know yourself will you know what is needed and how could so many people ever know themselves when science has changed their design from reaching that point.

Because people were connected to the Earth before through the matter of their DNA, people had a connection with the Earth, the frequency, the energy, the life force, the light of Earth, the nature, this is what nature is. This is what life is.

Most people have a problem with life and they're not ready for that conversation cause they're still fighting to defend the very thing that was designed to kill them.

The greater the suffering, the easier it is to physically close the minds of the people from living in a higher state from within.

Energy harnessing is used to draw people into cult followings, which triggers built in emotional responses, as the minds of the people have been simplified with negative lifestyles they cannot see past.

Because of the damage that has been done, anything beyond normalized living is too complex for coming into, leaving them with no spiritual insight, other than, their role as a social being.

Because people's light in higher thoughts can only be with the Earth, and because people are naturally connected to the Earth through their DNA, anything artificial that changes the DNA of the person, changes life.

When life is altered.

When the DNA is changed.

The genome has been modified, or cloned, and it is recast.

It has been genetically edited to become less than human.

The blood is life.

When life has been modernized the blood has been corrupted.

This breaks the connection with Earth.

People are no longer connected with the Earth as one in natural.

That's how they destroy life and call it progress.

They took humans away from the Earth and put them in an unnatural environment that is based on fear, destruction, violence, and negativity, that can only feed the strengths to darkness.

Once people are changed in a reversal, this starts to take away light from humans and the Earth, giving darkness more measure in power of changing the scales of energies that come to this planet.

One of the main evil workings behind this world's system is that it rewards misery and punishes any virtue of sovereign will.

The whole reason why people are falling further away from life, cause they do not carry a spiritual level of balanced elevation.

Humanity cannot reach its sole purpose by direction of tyrants.

You either die by their genocide or live on like a human being.

To save yourself you must clean yourself.

Clean does not just mean food and all other forms of lifestyle.

Clean ties directly to meaning as well.

A very big part of coming to know who you are and letting go of all things that have misled you.

If you are not focusing to change yourself to clean, then you are getting nowhere and will remain exactly where you are, as a part of this meaningless society.

This world is not the real home for humans and the people cannot base their home out of a physical structure from materialism.

Your one true home is within yourself while here on Earth. This allows the person to find happiness in any place as a home, no matter what they have or do not have.

True power comes from who you are and what you are doing from within.

What we choose to follow in our human experience determines you reaching the point of ascending the lower energies of modern living, as adoration for the world will corrupt your blood's frequency.

You cannot fight evil in the physical form and win.

Without a passage of light the body is apart from direction with identity.

The body and the soul must be together as one in spirit.

Only in the spiritual form of the person can evil be defeated.

A person's light inside with what they know is what makes them dangerous for the system.

Light has to come as an established connection between people.

Light is what exposes all forms of darkness and when the people have no light in them, they are consumed in darkness, only serving under everything it hides behind.

Remember what people have allowed to go on is what will continue and bring the world to a much darker place far beyond life.

Nothing ever with what they say or do is about saving lives.

Living in a higher standard of spiritual form is what saves lives from this world.

The light of the people is where true power comes from to defeat evil in all of it's delusive appearances and save multitudes of lives from this world.

Power is purpose. Purpose is in the soul.

The soul is the source of life in identity with purpose.

Living a life of purpose is the rise to greatness.

As there is light in purpose and light is the answer to life's purpose.

Unless you're living a life with true purpose, you're only wasting away, functioning in a life of meaningless appeals.

The presence of the soul will always differ from the physical mind.

You must recognize its existence.

Then act upon it.

One can shut out the other and vice versa. When subjugated by the mind, the material form will be overcome by the tendencies of the body. The person is then drawn into the path of the living, which whereby, desists spiritual linkage to the soul. That which presents the greater side of enlightenment past the compulsion to modern attachments.

Right now the people are involved with all the wrong people, places, and things.

Which are by design, to block them from having all the healthy and beautiful parts of life that are here for them.

Greatness only comes from beyond this world.

And how many can truly say before the universe that they are actually living to gain more than the world.

Your real power resides with who you are and what you're doing outside of the modern world.

When you're ready to leave the world behind your life will begin.

How far do you want to go with it and how much are you willing to invest.

When people want proof that light is here for them.

The light of day is the living experience that eternal life is before all.

You have to stand out in the light of life with who you are, not standing where the world tells you to be.

If you're not ready to lose everything in this world then you have nothing to gain outside of it.

Most of the world is afraid of change because they do not know any other way and that scares them to their core.

What humans don't know is what hurts them the most spiritually.

Yes it requires complete changes to your entire life and that means losing the system and a lot of people, places, and things from it. But after that you begin to have everything you need outside of this world coming into you. So that when you do touch back down into the world you have all the right people, places, and things, for you.

Chapter 7: The Science of Conspiracies
(The politics to everything)

Most people have no idea how the universe works because they are governed to the societies of this world, which is forecasted by modeled prediction through the art of design.

Do not let your illusions of what you believe life to be blind you into continually corrupting your blood.

All of modern reality manifests a false conscious belief system within the mind.

There is no accepting any part of this reality, as there is no consciousness to any part to it.

Society is molded around the ignorance of social engineering which gets you caught up in the normality of the world.

Your physical existence is a lie through living in a manufactured reality designed and built for control over life.

We live in a reality where people believe that truth is found on the internet and that is far from being a living source.

Truth can only be found through a true form of living.

The internet is not a source of truth.

It is artificial.

It is an artificial source of manufactured information written by algorithms.

Algorithms have become so powerful at how to control our mind's electrical mechanisms that they are already creating generative concepts that actuate a digital copy that overlays reality.

This is confusing people into believing false manifestations causing them to have less power of control over who they really are.

Over the years, I have destroyed all illusions of people's lives no matter who it was, or what they carried to value as being real.

Everything that I have written about and shown did not come from the internet.

It came from me and that's the difference people have no choice but to come to realize.

The problem is that people have been conditioned to view information as a conspiracy and this places the person in a reality with questioning outside of themselves as to whether something is real or in fact true.

Nothing they ever say or do makes sense.

People only know what they show them and what they show them is to manifest the science of conspiracies in their heads.

Conspiracies are not just politics.

They are the onset to every side of common logic, as the art of implementing confusion with what the people are thinking, saying, and carrying on with as their actions.

In Mesopotamia, they built a tower with advanced technology, which was used to confuse the people and everyone was speaking babel, or confusion as to what they are and what life really is.

This allowed Dracos to unite them under speaking a single language modeled after what they considered, "scientifically" altering the human being into a lesser form of this race.

Fast forward to the present and we are now there again.

Continual changing narratives are used in the elements of religion and state to implement the driving need for confusion.

Theories are ways to introduce scientific ideas to progress modern thinking and carry no relevance to what is real.

When I wrote they are masters of mind control, a lot of people took that as a grain of sand without having any acknowledgement to understanding the degree of seriousness.

It's another level of alchemical processes that biologically recodes the blood of the human psyche through existing for the moment by scientific reasoning: "The truth can be whatever you want it to be. The truth is whatever you make it to be."

The basis of confusion to driving the mystic ideology behind the aim of infinite possibilities and outcomes.

Driving the people farther from understanding what they are and even to the extremes of the consideration that they are not human, but rather, are something else outside of being human.

They promote the science of being whoever you want to be, to stop you from becoming who you need to be.

That's why they start off at the earliest of age in the schools telling the children "What they want to be when they grow up."

And they are now using 5g based electromagnetics to gain entry, as a direct portal of channeling into the deepest regions within the mind, to control the creation of light within thought control processes.

You can do anything you want.

You can be anything you want outside of who you need to be and what you should be doing.

The theoretical state to the infinite parallels in a multiverse reality to reaching the state of god or quantum machine intelligence.

The mark of influence involves electrical passage into the mind.

Linking the mind as giving creation to a new technological Genesis.

One in which people have been willingly helping to recreate as the image of God.

A new world of human converts created in connection to the central mind of this God as a single living biological consciousness.

As they have built the new world into the digital image, reality has followed with the death of nature passing.

The corruption of the human being sparked the death of creation.

Once the people succumbed to the force of this evil their transformation began into becoming it.

The science behind the aim of the church to promote people to a state of material form rather than spiritual form.

This created a society of humans that are more concerned with having than with being.

Remember Draco is not spiritual.

They have no connection to the light of life.

Everything they devise as said "spiritual" is fashioned through black energy practices as "false miracles" presented to humans as to what life is.

The soul of the human is light which interacts with waves of energy transmitting information through the DNA, powering the physical body with knowledge as existence. When the person is disconnected with no state of spirit, that light has been taken from them, the blood is poisoned, and they are in death having no connection with any type of ascending involvement.

They have nothing in the spiritual with eternal life, only everything in the physical existent. That's why nothing in the material world matters.

The current state of the world is what you get when you have mass populations who follow the narrative of society.

Look at how far down they have brought the people that they don't even know what it means to be human anymore.

The way the people are living they have no compassion, or desire, to be responsive as sentient beings.

Only they function as lifeless vessels, alive in a bubble of repetitive make-believe, which they hold to core value as being real.

People have been modified as fake inside as everything they surround themselves with on the outside.

This is why they have no understanding of what is real.

How could anyone know anything when they have everyone paying attention to everything outside of themselves.

You're seeing what they want you to see, not what you need to see!

This is not where you find the proof that you're looking for that you need.

People have to go beyond the conspiracy of what they have been shown life as to be.

Science is what people are trusting as their given source, in searching for what they consider real proof. Science is what people follow, who do not have spiritual perspective to know, that, science comes from the darkest of lies.

Science became human consciousness, replacing individual thinking with technological interpretation.

You are living in a physical nightmare that has gradually become far worse, contributing to turning you into a residual version of yourself.

A mental projection of your mind as a digital consciousness.

Technology has created the tools for destabilizing and eroding the very fabric of society in every country worldwide.

People believe that the internet and social media are powerful for bringing change.

In reality, both are nothing more than a tool being used for leading people on, to follow the creation to their own destructive revolution.

Most conspiracy groups, activists, and social media influencers put out information to promote engagements, which looks good from a clickbait standpoint, but popularity numbers get you nowhere, other than producing followers who are still choosing to power some form of the system with their life force.

Real truth remains within to be discovered and they have the vast majority of the world following the political shows of conspiracies, thinking they are in some kind of awakened journey.

They promote the age of information as a revolution to change built around swift political shifts being implemented all over the globe through events shaped by sinister leftists and radical nationalist forces insinuating violent uprisings and progressive aims popularized as "anarchy".

The word "anarchy" carries strong symbolic meaning to the power it casts over society as its vibration spells heavy runic energy into the minds of humans.

(Incidentally the word "anarchy comes from the cuneiform sign for "An" meaning "Dingir" which was the God of the Babylonian pantheon.

"Archy" means "rule".

Symbolically the word "anarchy" translates from ancient text to mean "God's rule".)

Thinking and knowing are two completely different experiences.

One is highly unique.

The other is a common base in civilized logic.

Because conspiracies are about you investing the energy of matter into everything that's manifested within the delusions of modern thinking.

People who speak and write conspiracy are not knowledged. Conspiracy and knowledge are not the same.

Conspiracy is what's promoted in a digital reality where nothing you see and hear is real.

There are politics to everything in directing thought and to control thought is corruption to the blood.

That is exactly why conspiracy groups and individuals who call themselves that title were created.

To lead everyone down the path of where they want you to be as viewed by society.

Conspiracy groups and anybody who follows them are considered to be dangerous and mentally unfit.

None of the people involved with any type of activist conspiracy group actually know anything beneficial, other than some baseless information they immersed in off the internet.

There is no questioning everything.

There is only knowing.

Questioning everything only fuels the conspiracy of debate in division and conspiracies have nothing to do with truth.

Debates and division create doubts, denial, negativity, and ignorance, which manifests a corruption of mental disease in the mind.

They will debate with others and discuss information of different sides to try and find some sort of truth for them to follow but it does not work like that.

Debates and discussions are nothing more than scientific reasoning and there is no backing the lies of science.

There is no debating the ignorance of scientific reasoning.

Everything in life is based around the mental arts of manipulative energies used to make you blind to what the real source of the problems is.

People don't want to wake up to knowing cause they cannot mentally process their illusions of life being based from evil.

Those who need scientific proof in order to shape their beliefs have been molded in the vagueness of modernization.

The source of this world's deviation from light.

Those who can never handle it always proceed elsewhere, carrying on with their meaningless lives to choose their side of poison leading the death of their soul.

There is no choosing sides.

When you choose a side, you're still very much a part of the system, making you part of the problem with why this world is the way that it is.

They control 360 degrees to all sides with everything and they play humans back and forth from every angle within it.

There are different factions of them from the same bloodline who are working together through the center stage of modeling a public image of extreme opposition.

Politics are the visuals of fake opposition.

Conflicting bureaucratic views are used as practices of divisional concepts using identity politics, which are strong tactics to divert attention away from the real issues with who is standing before you destroying the living.

Conspiracies are the energy of their direction and have nothing to do with truth.

Truth cannot be given by one person to another.

Truth is a living experience that happens throughout certain measures of the life span as cycled from within.

This is part of a complete life changing journey from leaving out of the system.

Truth is not easy to find.

If you do find it within yourself, you have learned that it does not conform to the levels of your personal comfort zone with what you have been in part to your entire life before.

Truth is a painful experience.

It will take everything from you in this world.

But it also gives you everything outside of it to come back into after you are ready to live in the light of your newfound purpose.

Because truth is light and they do not come from light.

Do you understand?

Conspiracies come from politics, which have nothing to do with truth other than to mock that which they make to happen.

The only conspiracy is the one they put inside your head to keep you searching everywhere outside of yourself for the proof to know what is real within yourself.

Their goals are achieved by submission to freedom.

The people are forced to submit to the declarations of freedom.

Freedom is an elusive concept rife with individual meaning.

There is strength in surrender, bringing unity, as long as the people are willing to remain slaves to their freedoms.

Their main goal is to destroy the people's right to freewill on this planet.

Freewill is the birth to life with meaning and purpose outside of being a slave through your own minds choosing.

The more the people listen to the progressive tyranny the more power they have, the closer they are to reaching that goal.

Once they have done this they have reached full destructive control over all life.

The left and the right are two opposing sides that lead to the same evil.

The right accuses the left of that which they are also guilty of as well.

The right was used to misguide people going into the age of truth starting in 2012 by turning people into conspiracy fundamentalists, who are devoted to backing the technocrats and conservative idealists carrying on the criminal elements behind fascist ultra-nationalism.

In order to keep people trapped from leaving the system, they use the right to facilitate conspiracy, in order to own tens of millions who are factionalized behind the shepherds who were placed to stand before them.

They publicize what they are doing through the political right.

This is the right hand path which represents the illusion of choice with systemic human morals and values.

Conspiracy politics is a tool of patriotic rhetoric used to reaffirm the dominant and established power of nationalist groups.

They put it right in everybody's face what they are doing in destroying the human race, but nobody sees this cause they are too busy focusing their life force into choosing a side to fight for, within the choices they are presented.

If you support law enforcement, military, healthcare, education, chasing money for material lifestyles and have some kind of faith in the overall system in general, you are not in control of your own existence.

Because you are still living under the illusions of life that you are actually free, when you are not at all free. It is the complete opposite. You are just a slave, fighting with life, in a manufactured reality built for control.

Conspiracies are political media based on the physical realm outside of self knowledge.

Conspiracism is the state of belief with questioning everything outside of the soul.

Questioning everything is a conspiracy based sense of thinking with searching outside of yourself for the answers.

Stop asking questions and pay attention. Make daily changes.

Stay on the path and strictly focus on those changes for yourself.

You need to stay consistent with these changes, because, falling back into chasing negative habits is all too familiar within the general public.

Society was engineered to break you down as a person and mold you into who it wants you to be.

So without any part of society do you even know who you are as a person without it?

Illusions of life are everywhere all around you. Your home. Your money.

Your job you report to every day, takes power out of your soul, so that you remain spiritually broken with no time to realize a change is needed.

Everything you know as life is the world that has been created to hide the truth of life from you.

Modern education is indoctrinated into the baseless mindset of following cultural society.

The evolution of culture society.

People have no understanding cause their blood has been changed to follow the narrative of this basis.

Greater knowledge has always baffled primitive men.

The fallout of society will always approach you with the narrative of normalization, as modern minds carry on in the profane disbeliefs of civilized logic.

Most people speak from a standpoint of wanting to be politically correct with what they side with.

A baseless conversation for the dumbed down, who live in a bubble of belief, that all should have a just and equal fair chance in their servitude as slaves.

The people will always have something to try and prove in the physical to compensate for what they do not have with a spiritual life.

It does not matter what society thinks as a whole.

What society chooses to believe is irrelevant to the truth.

The truth should always be revealed no matter who it offends.

It is not meant for everybody, but it will reach those who it is designed for.

And that's what matters the most.

Society is not the people and people are not society.

If you think that any part of functioning in society is real, go bang your head off a brick wall and get a clue.

Society as you know it, is a joke, and there is no accepting any part of this reality as truth.

Society is the illusion that has led the people to believe what they are.

They worship everything in the physical as the intangible senses of the mind.

Society is a lost cause and we do not have moments to waste on those who cannot experience an understanding of what is needed.

Nobody wants to listen cause they are too busy feeling low from their physical connection with how society has gone.

They let the system get their emotions with mental control and that's exactly what it was designed to do.

When beliefs are led by emotions this places the person outside of light revealing consciousness.

Emotions reside from the subconscious mind, which no modern person has any control over, but yet, controls close to 100 percent of their daily thoughts.

You cannot focus your life around the design of the system.

The system is always to take the life out of the people.

You have to discover the path of who you really are outside of who society influences you to be.

People really need to get to a better point of focusing on going deeper with what they think they know, for having a much better understanding of true reality.

This is the only way you will see truth in the right form for the mind to develop real consciousness.

I'm not here to pamper people with fantasies in conspiracy literature.

I'm here to bring people to a breaking point in their lives with making vital life changes to reach a higher state of existence.

There is no in between comfort zone, where you can sit idle in your daily living, thinking that everything is going to be ok in this reality.

You can keep making excuses and siding to defend your illusions of life, but it's not going to change any of what is now taking place here.

A harsh awakening but it's necessary, nobody ever said the journey was going to be easy.

Your illusions of life do not matter.

What matters is what you are doing for yourself to counter what they are doing to you. I'm not here to show you proof to make you believe anything.

I am only here to show you some things to open the doorway in your mind for you to walk through and make the necessary changes to begin the path to a life undreamed.

Finding purpose with life is living the dream. When you find your purpose in life you will never suffer another day ever again.

You need to find the proof in yourself that is the only way the eyes can be opened to see life's true light.

9 781964 630304